iSplash-I: High Performance Swimming Motion of a Carangiform Robotic Fish with Full-Body Coordination

By
Richard James Clapham

iSplash Robotics

Published by *iSplash* Robotics
www.isplash-robotics.com
Copyright © 2016

Published by *iSplash* Robotics 2016
Illustration by Richard James Clapham PhD

All rights reserved. No part of this publication may be reproduced, stored in a retrieval system or transmitted in any form or by any means, electronic, mechanical, photocopying or otherwise without the prior permission of *iSplash* Robotics.

ISBN-13: 978-1537259635
ISBN-10: 1537259636

Thank you for your purchase.

***Abstract*—**This book presents a novel robotic fish, *iSplash*-I, with full-body coordination and high performance carangiform swimming motion. The proposed full-body length swimming motion coordinates anterior, mid-body and posterior displacements in an attempt to reduce the large kinematic errors in the existing free swimming robotic fish. It optimizes forces around the center of mass and initiates the starting moment of added mass upstream. A novel mechanical drive system was devised operating in the two swimming patterns. Experimental results show, that the proposed carangiform swimming motion approach has significantly outperformed the traditional posterior confined undulatory swimming pattern approach in terms of the speed measured in body lengths/ second, achieving a maximum velocity of 3.4BL/s and consistently generating a velocity of 2.8BL/s at 6.6Hz.

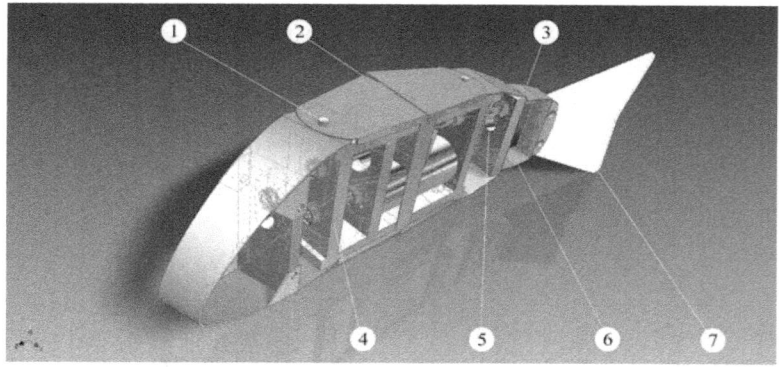

Figure 1. *iSplash*-I: 1-Anterior actuation; 2-Midbody; 3-Thick peduncle; 4-Transmission system; 5-Driven tail plate; 6-Tendons; 7-Compliant fin.

I. INTRODUCTION

Underwater exploration is a physically demanding task. A robotic vehicle must navigate in an unpredictable surrounding environment which includes many disturbances and non-uniformities due to the transient changes exerted from the liquid. These physical forces applied to the underwater vehicle (UV) can lead to inaccurate navigation and even failure of the operation as traditional propeller driven systems are unable to adapt to the dynamic environment. Some operational examples are military surveillance, mine countermeasure, inspection and pollution mapping. UV's have particularly high cost of transport during low speed mobility [1]. This is where fish excel, generating large transient forces efficiently by coordinating their body motion [2],[3]. In addition, adversely the dynamic environment can be advantageous to fish locomotion, as they demonstrate the ability to extract energy from the upstream vortices [4]. Therefore there is great potential to improve UV's by imitating the swimming fish with the highest locomotive performance operating within the desired environment and Reynolds numbers. Although there are still many aspects of locomotion to address, linear propulsion could be considered the greatest challenge.

Most work in biomimetic underwater propulsion has focused on hydrodynamic mechanisms. Some examples of novel design approaches and their maximum speeds are Barrett's hyper-redundant parameterized Robotuna which achieved a velocity of 0.65 body lengths/ second (BL/s) (0.7m/s) [5], Yu's optimized discrete assembly prototype achieving a velocity of 0.8BL/s (0.32m/s) [6], Liu's G9 Carangiform swimmer achieving a velocity of 1.02BL/s (0.5m/s) [7] and Valdivia y Alvarado's compliant structure assembly achieving a velocity of 1.1BL/s (0.32m/s) [8]. Currently the low speeds of robotic carangiform swimmers are unpractical for operation, peaking at speeds of 1Bl/s. In comparison, transient speeds of comparable live fish averaging at 10BL/s have been measured by Bainbridge [9].

Reproducing the propulsive force of fish is a complex challenge. The development of a novel mechanism must take a number of factors into consideration, including morphological properties, weight distribution, efficient transmission principles, power density constraints and kinematic parameters. In particular accurately replicating the linear swimming motion has proven to be difficult and free swimming robotic fish have significant kinematic parameter errors. The lateral and thrust forces are not optimized and as a consequence excessive anterior destabilization in the yaw plane due to the concentration of posterior thrust creates reaction forces around the centre of mass. In turn the anterior creates posterior displacement errors. As a result the body wave motion along the full length of body has large matching errors in comparison to the swimming patterns of live fish leading to reduced performance and high cost of transport.

This research project considered the factors contributing to the low hydrodynamic performance of current robotic fish within linear locomotion and proposed four main objectives: (i) Introduce a new swimming pattern to reduce the kinematic parameter errors by coordinating transverse displacements along the body length. (ii) Allow for efficient energy transfer by engineering a mechanism that takes into account hardware and material constraints so that propulsion is not restricted. (iii) Develop a prototype to improve stability in the vertical and specifically the horizontal plane, by optimizing the lateral and thrust forces around the center of mass. (iv) Validate the proposed swimming motion by realizing a mechanism capable of consistent free swimming operation, measuring its achievement in terms of speed, thrust, and energy consumption over a range of frequencies.

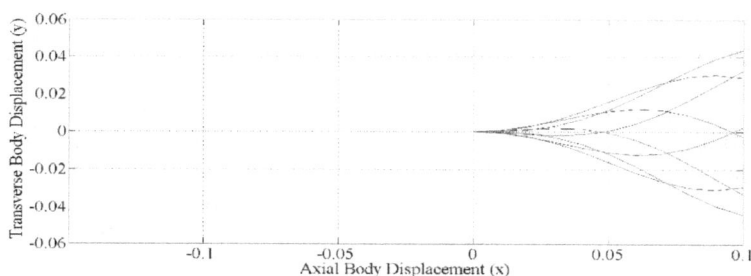

Figure 2. Mode 1: Wave form is confined to the posterior 2/5. Parameters have been determined from experimental tests.

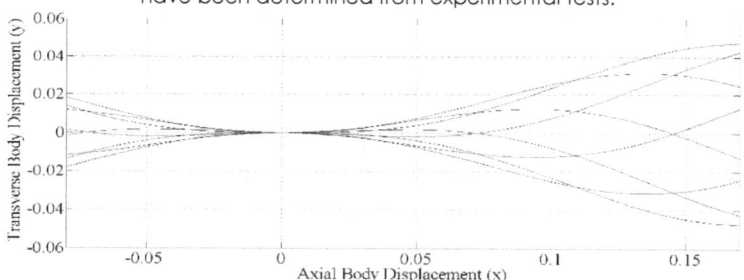

Figure 3. Mode 2: Full-body coordination. The kinematic parameters have been determined from experimental tests.

The remainder of the paper is organized as follows. Section II presents the traditional approach of robotic fish and introduces a new full-body length swimming motion approach. Section III describes the construction method of a novel robotic fish, *iSplash*-I. Section IV describes the field trials undertaken and the experimental results obtained. Concluding remarks and future work are given in Section V.

II. Design Methodology

A. Traditional Approach

Modeling from body and/or caudal fin (BCF) swimmers, the selected carangiform swimming mode can be identified by the wave length and amplitude envelope. The Cyprinus carpio (common carp) has been chosen specifically for its high locomotive performance [10].

We only consider modeling within the confinements of the horizontal plane where the kinematics of propulsion are commonly reduced to the form of a traveling wave, concentrated to the posterior, varying in amplitude along the length, smoothly increasing towards the tail [2]. Present robotic swimmers adopted this method which limits undulatory motions, typically to <1/2 the body length towards the posterior and the wave form motion consists of one positive phase and one negative phase. The commonly adopted model proposed in [5], is in the form of:

$$y_{body}(x,t) = (c_1 x + c_2 x^2)\sin(kx + \omega t) \qquad (1)$$

where y_{body} is the transverse displacement of the body; x is the displacement along the main axis starting from the nose of the robotic fish; $k = 2\pi/\lambda$ is the wave number; λ is the body wave length; $\omega = 2\pi f$ is the body wave frequency; c_1 is the linear wave amplitude envelope and c_2 is the quadratic wave. The parameters $P = \{c_1, c_2, k, \omega\}$ can be adjusted to achieve the desired posterior swimming pattern.

B. Proposed Full-Body Swimming Motion

Propulsion of carangiform swimming is associated with the method of added-mass [11]. Each propulsive segment of the travelling wave creates a force against the surrounding water generating momentum. This causes a reaction force F_R from water onto the propulsive segment. F_R normal to the propulsive segment is decomposed into the lateral F_L component which can lead to energy loss and anterior destabilization and the thrust F_T component providing propulsion increasing in magnitude towards the tail. The overall magnitude of added-mass passing downstream is approximately measured as the water mass accelerated and its acceleration.

Therefore it is proposed that initiating the starting moment of added mass upstream and optimizing the F_L and F_T forces around the center of mass would increase the overall magnitude of thrust contributing to increased forward velocity. In consideration of this, we designed a novel robotic fish which can operate in two swimming patterns: (i) Applying a traditional rigid mid-body and anterior. Concentrating the undulations and degrees of freedom (DOF) to the posterior end of the body length which will be described as Mode 1, illustrated in Fig. 2; (ii) Based on intensive observation and fluid flow assumptions a new full-body carangiform swimming pattern is introduced. The coordination of anterior, mid-body and posterior body motions are proposed in an attempt to reduce kinematic parameter errors, this will be described as Mode 2, illustrated in Fig. 3.

The models midline and body motion parameters were first established based on observation and published data from literature providing an initial engineering reference. The wave form motion first developed for a discrete rigid anterior prototype in [7] can be extended to represent the full body motions of Mode 2 in the form:

$$y_{\text{body}}(x,t) = \left(c_1 x + c_2 x^2\right)\sin\left(kx + \omega t\right) - c_1 x \sin\left(\omega t\right) \quad (2)$$

The relationships between the defined parameters P = {0.44,0,21.6,8} shown in Fig. 3 can first be found by evaluating the x location pivot at 0. In the kinematic pattern of Mode 2, the fraction of body length displaced is equal to the anguilliform swimming mode but reflects changes in the wave form. The anguilliform swim pattern is defined by large amplitude undulations propagating from nose to tail. The newly introduced Mode 2 applies an oscillatory motion to head and mid-body and pivots the entire body around a single point associated with the carangiform fish swimming motion [2].

Although this is the first account of applying full-body actuation to a research prototype fish, mechanisms such as "vortex peg" and "undulating pump" and flow visualization techniques have been proposed [12][13] from published biological studies, indicating a possible fluid body interaction that contributes to propulsive thrust is generated upstream to the posterior section. The muscle activity in the anterior has been measured to be low, suggesting that accurate modeling of the kinematics could be more significant than anterior force in improving energy transfer.

As previously mentioned, anterior destabilization has been difficult to control [6][7][8], as passive rigid anterior mechanisms recoil around the center of mass. Free swimming robotic fish have excessive head swing, similar in magnitude to the posterior which increases drag. The proposed Mode 2 drives the anterior into the unwanted yaw direction, in an attempt to reduce amplitude errors by optimizing the F_R around the center of mass. It has also been noted in [2], that the morphological adaptations of reduced depth at the peduncle, increased depth of body towards the anterior and vertical compression minimize recoil forces.

III. New Construction Method

A. Mechanical Design

Mechanical structure limitations set a great challenge when modeling the displacements within the travelling wave. Current methods typically adopt either a discrete assembly [6], [7] or compliant structure [8] but both are seen to have limitations. A construction method using structural compliance combined with a rigid discrete assembly is proposed. The arrangement distributes 3 degrees of freedom (DOF) and 1 passive DOF along the axial length shown in Fig. 4. Mode 1 disregards transverse displacements of links I, II, III whereas Mode 2 actuates all DOFs along the axial length to provide anterior and mid-body transversal displacements.

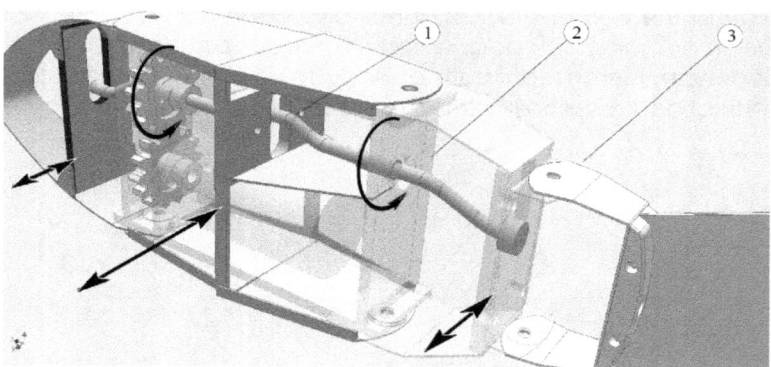

Figure 5. Power transmission system: 1-Transision plate; 2-Crankshaft; 3-Free end of link and connecting pivot.

The development allowed for both Modes of operation to be applied to the same prototype by adjusting the configuration. Uniform material properties were chosen for links I-III and stiffness distribution begins at DOF 3 and continues to the tail tip. To provide the undulatory motion a compliant caudal fin is attached to the link V and is actuated by tendons anchored to the main housing rear bulkhead. The developed mechanism allows for the expansion of the tendons and material stiffness of the caudal fin to be adjusted experimentally to provide the targeted curves during free swimming at various frequencies.

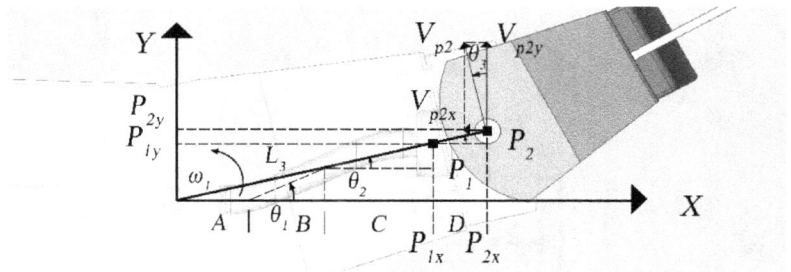

Figure 6. Schematic drawing of the tail offset drive crank and linkage;

The approximation of a traveling wave using joints I-V and turning angles of DOF 1-4 are shown in Fig. 4. Details of the fully discretized body wave fitting method are given in [6], [7]. The location of joints in the series can be determined by parameterized fitment to a spatial and time dependent body wave. The discrete construction method can be defined as a series of links or N links. N being the number of links after DOF 1 typically <6 due to structural limitations, more links reduce curve alignment errors. The aim of the design is to improve complexity of motion without an increase of structural parts. The link end points are shown in Fig. 4, it can be seen that the arrangement of DOF's distributed along the length of the body provides an accurate curve alignment reducing large errors and excrescences in the outer profile. In addition we have observed that the aerofoil section NACA (12)520 based on camber, chord and thickness can be utilized to illustrate the outer structure profile of Mode 2 which we propose contributes to the fluid flow interaction.

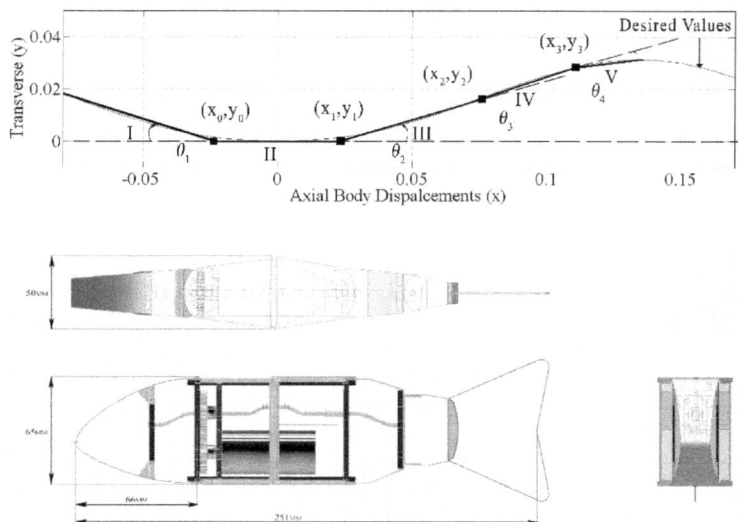

Figure 4. Link approximation (top); 1-Plan; 2-Profile; 3-Front (bottom)

B. Power Transmission System

The developed transmission system providing rotary power to linear oscillations is illustrated in Figs. 5 and 6. All actuated links are directly driven by the five bearing crank shaft providing an equal power distribution. The developed mechanical design required high-precision engineering of the chassis and crankshaft to avoid deadlock and reduce friction. The driven link amplitudes are determined by the offset cranks, L_3 represents one of the discrete links of the structure. The maximum amplitude of the link length L_3 at point P_2 is determined by the predetermined maximum crank offset P_1. The coordinates of P_1 (P_{1x}, P_{1y}) and P_2 (P_{2x}, P_{2y}) can be derived by:

$$\begin{cases} P_{1x} = A + B\cos\theta_1 + C\cos\theta_2 \\ P_{1y} = B\sin\theta_1 + C\sin\theta_2 \end{cases} \begin{cases} P_{2x} = P_{1x} + D\sin\theta_2 \\ P_{2y} = P_{1y} + D\sin\theta_2 \end{cases} \quad (3)$$

The length of L_3 can be derived by $L_3^2 = P_{2x}^2 + P_{2y}^2$. Assume that ω_1 is the angular velocity of the link L_3, and the velocity vector V_{P2} is perpendicular to L_3. We have:

$$\begin{cases} V_{p2x} = -\omega_1 L_3 \sin\theta_3 \\ V_{p2y} = \omega_1 L_3 \cos\theta_3 \end{cases} \quad (4)$$

where V_{p2x} and V_{p2y} are the decomposed vectors of the velocity vector $V_{P2} = \omega_1 L_3$.

C. Fabrication

iSplash-I shown in Fig. 7 was engineered as a morphological approximation of the common carp. The physical specifications are given in Table I. We devised a structurally robust prototype allowing for consistency of operation at high frequencies, as force has to be applied to the water and reactively, the opposing force is applied to the vehicle. All structural parts were precision engineered, hand fitted and assembled. A consideration of the development took hardware and material constraints into account, so that geometric and kinematic parameters are not affected. The hydro-static streamlined profile was optimized by favorably positioning the maximum thickness of the cross section, reducing pressure drag. In [10], the cross section has been measured to be optimal at 0.2 of the body length. These aspects relate to amount of resistance during forward motion and were taken into consideration within the design.

Increasing endurance is a desirable feature of a UV. Current robotic fish are still limited to short operational times as energy losses can be produced in many stages of the mechanical transfer. Recent designs have found it is advantageous to utilize a single electrical motor for actuation [8]. The classical actuator is still the most effective way of providing power at high frequencies and reduces energy consumption over multilink discrete assemblies. Mass and volume distribution are key principles of stability in the horizontal and vertical planes. A single actuator power transmission system can be positioned in the optimum location. In contrast multilink servo assemblies are limited as mass and volume are confined to the posterior.

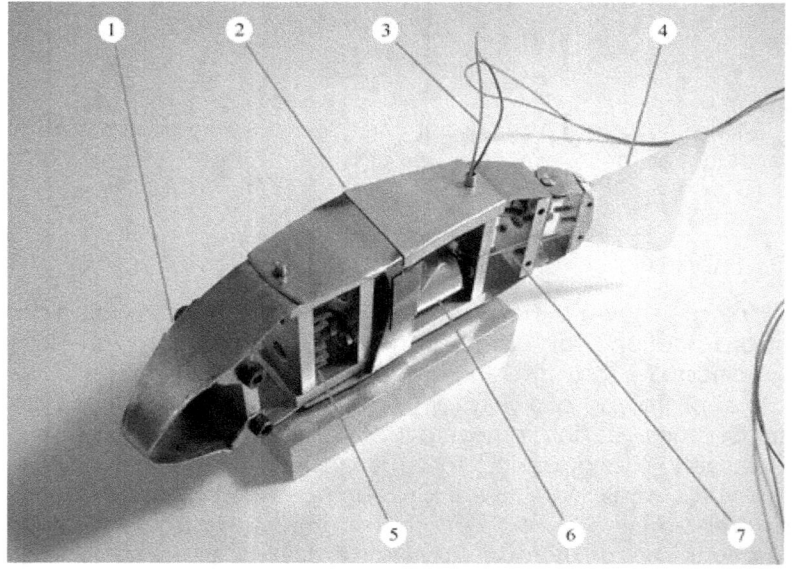

Figure 7. Inner Structure of *iSplash*: 1-Mild steel space frame; 2-Mid-body driven plate; 3-External source cables; 4-Polypropylene caudal fin; 5-Aluminium main bulkhead; 6-electric motor; 7-Offset crank shaft;

The body has open loop stability if the relative position of buoyancy is higher than the center of mass as the surrounding fluid counterbalances the gravitational weight [14]. Therefore the hydrostatic buoyancy level and stability were solved by adjusting material properties and configuration. Stability was found to be particularly difficult to maintain during free swimming at high frequencies.

Lastly, the inner structure of the prototype is negatively buoyant. A significant development of the prototype was a watertight skin that allowed unrestricted flexing of the external surface and provided the volume needed to maintain neutral buoyancy.

TABLE I. PHYSICAL PARAMETERS OF *iSPLASH*-I

Parameter	Specific Value
Body Size: m	0.25 x 0.05x 0.0 62
Body Mass : Kg	0.367
Maxium Velocity: BL/s (m/s)	3.4 (0.88)
Noload Maximum Frequency: Hz	8
Actuator	Single Electric motor
Power Supply	12V Pb External Battery Supply
Fabrication	Low Tolerance Engineering
Materials	Aluminum, Mild Steel, Stainless
Swimming Mode	Linear Locomotion
Tail Material	Polypropylene
Skin Material	Polypropylene
Thickness of Caudal Fin : mm	1
Caudal Fin Aspect Ratio: AR	1.73

IV. Experimental Procedure and Results

A. Field Trials

A series of experiments were undertaken in order to verify the proposed swimming pattern by assessing the locomotive performance of Modes 1 and 2 in terms of speed, thrust and energy consumption at frequencies in the range of 2-8Hz. Experiments were conducted within a 1m long x 0.5m wide x 0.25m deep test tank. Stabilized free swimming over a distance of 0.5m was used to measure speed with a 0.5m acceleration distance. The prototype had sufficient space to move without disturbances from side boundaries and the free surface. Measurements were averaged over many cycles once consistency of operation was achieved and steady state swimming was obtained.

Although the prototype was measured to have a higher mechanical efficiency with an oiled filled structure, the developed skin proved inconsistent. Therefore, all runs were completed actuating the prototype with a water filled structure. This method attained consistency of operation providing stabilized swimming and maintaining the required buoyancy within the depth of the testing tank, whilst gently skimming the bottom surface. Velocity greatly reduced during runs when the skin detached, the build became negatively buoyant, destabilized or the cross-sectional area was increased.

B. Swimming Pattern Observation

Fig. 8 shows snapshots of Mode 2 in eight instances with time intervals of 0.02s throughout one body cycle. The midline was tracked at 50 frames per second to provide the amplitude envelopes of the anterior and posterior for comparison. Good agreement with fish kinematic data is a difficult task and current free swimming robotic fish have shown excessive head and tail amplitude errors. When comparing Modes 1 and 2, Mode 2 was found to reduce the head amplitude by over half from 0.17 (0.044m) of the body length in Mode 1 to 0.07

(0.018m). The tail amplitude of the common carp is 0.1 [9] [10], larger values were found to increase performance. Both Modes 1 and 2 were able to attain amplitudes of 0.17 (0.044m). The location of the midline pivot point should be in the range of 0.15-0.25 of the body length [10]. Mode 2 has a reduced error location of 0.33 in comparison to 0.5 in Mode 1. Indicating Mode 2 significantly reduces matching errors.

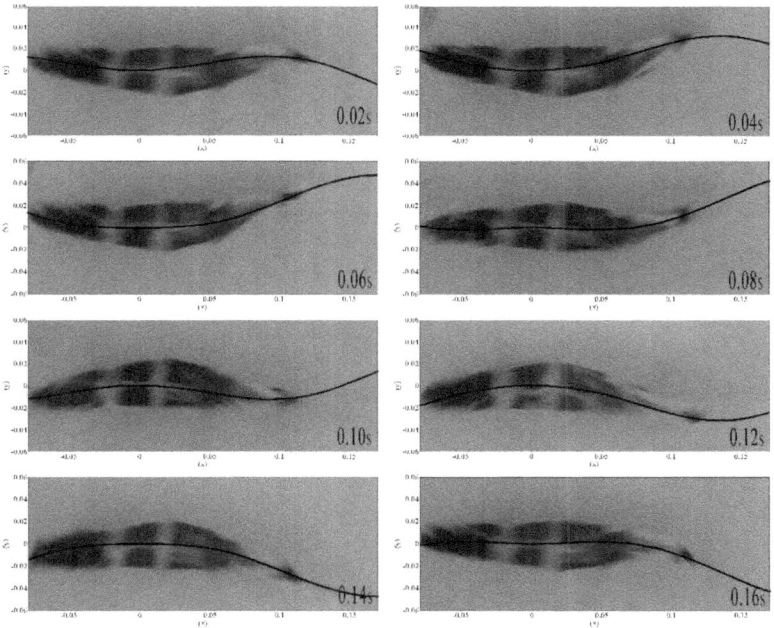

Figure 8. Mode 2 during one body cycle, eight instances every 0.02s.

In addition it was observed that the posterior 2/5 of the body length deforms due to stiffness distribution providing a smooth transition phase between body and tail. Although high aspect ratio (AR) caudal fins have been found to produce greater efficiency [4], in initial testing a low aspect ratio tail provided higher speeds. AR is calculated using: AR=b2/Sc where b squared is the fin span and Sc is the projected fin area. AR in this case was 1.73.

TABLE II. COMPARISON OF TEST RESULTS BETWEEN MODES 1 & 2

Parameters	Mode 1	Mode 2
Reynold Number: Re (10^5)	1.4	1.7
Shrouhal Number: St	0.48	0.41
Maxium Thrust: N	0.63	1.17
Consistant Maxium Velocity: BL/s (m/s)	2.2 (0.55)	2.8 (0.70)
Frequency: Hz	6.1	6.6
Max Power Comsumption Air: W	3.48	3.76
Max Power Comsumption Water: W	5.76	7.68
Swimming Number: Sw	0.36	0.42
Head Swing Amplitude: m	0.044	0.018
Tail Swing Amplitude: m	0.044	0.044
Test Run Distance: m	0.5	0.5

C. Experimental Results

Fig. 9 shows the average energy economy in relationship to driven frequency, comparing both Modes in air and water. This comparison measured the value of the increased resistance during locomotion due to the surrounding liquid. The measuring of the energy economy and thrust took many cycles to average, as the swimming motion produces fluctuating readings within a single body motion cycle. Both Modes actuating in water resulted in an increase in energy consumption, i.e. Mode 2 increasing from 3.76W to 7.68W and Mode 1 increasing from 3.48W to 5.76W.

As the configurations of robotic fish show various hardware and morphological properties, the main value of comparison has become speed divided by body length (BL/s). In this case the body length is measured from nose tip to the most posterior extremity of the tail.

The relationship between velocity and driven frequency is shown in Fig. 11. The corresponding values of Modes 1 and 2 during consistent swimming were measured and compared to current robotic fish. Mode 1 achieved maximum velocity of 2.2Bl/s (0.55m/s), at 6.1Hz. Mode 2 increased maximum velocity to 2.8BL/s (0.70m/s) at 6.6Hz. Mode 2 has significantly increased performance in comparison with current robotic fish which typically peak around 1BL/s. An initial value of 3.4BL/s (0.87m/s) at 6.8Hz was recorded by Mode 2 with an oil filled structure. Sealing the developed skin when in contact with oil could not be maintained and skin detachment consistently affected stability and buoyancy, greatly reducing performance.

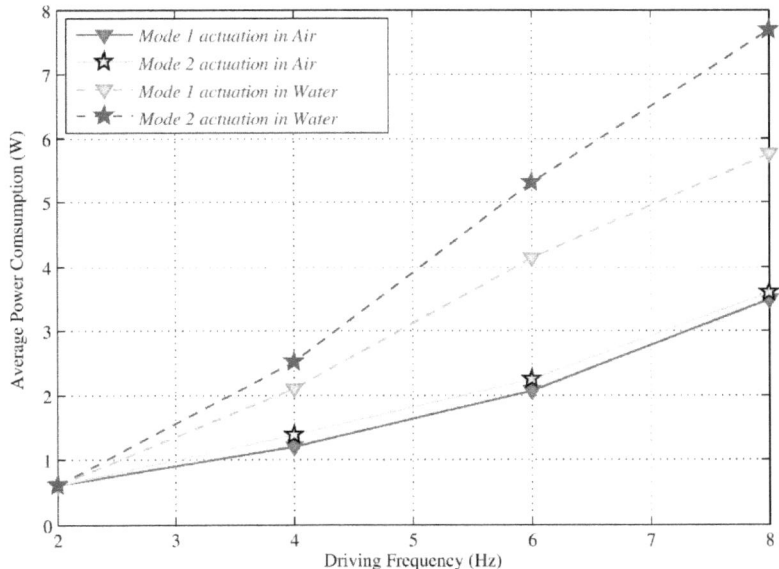

Figure 9. Comparison of average electrical power consumption over driven frequency of both Modes, actuating in air and water.

We can notice that Mode 2 had an 85% increase of thrust (Fig. 10) and a 27% increase in velocity over Mode 1 whilst consuming only 7.68W of power at 2.8BL/s. As the power supply contributes to a significant portion of the total mass, high energy efficiency is important. The measured low energy consumption indicates that the next generation could carry its own power supply within a comparable geometric frame with good endurance.

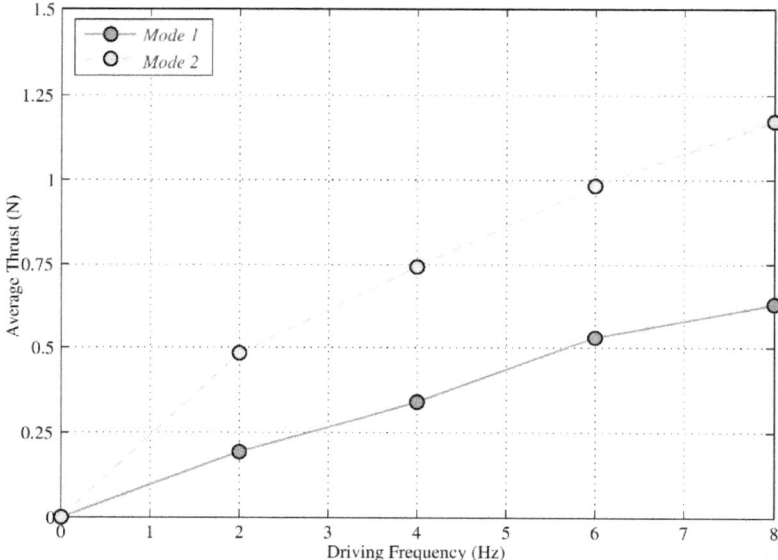

Figure 10. Comparison of average thrust in relationship to driven frequency achieved by both Modes.

A prominent parameter for analyzing BCF locomotive performance is the Strouhal number (St), defined as $St=fA/U$, where f denotes the frequency, A denotes the tail amplitude and U is the average forward velocity. St is considered optimal within the range of $0.25 < St < 0.40$. Mode 1 has a peak St of 0.48 under the condition of $Re = 1.4 \times 10^5$ and Mode 2 consistently measured a $St = 0.41$ and peaked at a $St = 0.34$ under a condition of $Re = 2.2 \times 10^5$. A comparable live fish was measured in [3], to have a $St = 0.34$, $Re = 2$~8×10^5. Applying Mode 2 shows a high performance increase within the St optimal range and achieves the higher cruising speeds of swimming fish.

A significant relationship between velocity and driven frequency was found. As higher frequencies were applied velocity increased in both Modes, matching the reported findings of live fish [9]. From this it can be assumed that a further increase of frequency applied to this prototype may continue to increase its performance.

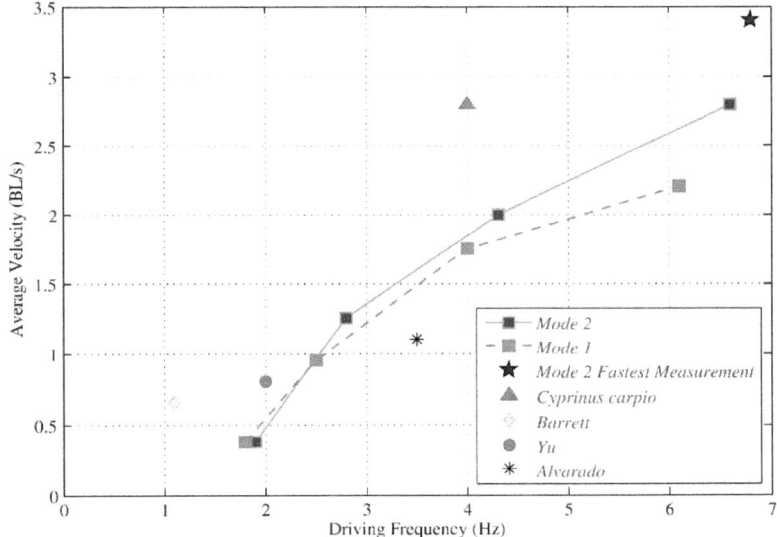

Figure 11. Comparison of average velocities achieved by both Modes, contrasted against current robotic fish and the common carp.

V. Conclusion and Future Work

In this paper we show conclusively that by coordinating the full-body length of the carangiform swimming motion a significant increase in performance, in terms of linear swimming speed, is gained over the traditional posterior confined wave form. The innovative mechanical drive system increased maximum velocity over the current published robotic fish. In fact, the robotic prototype can continue to increase its velocity as increased frequencies were applied, indicating, that the high swimming speeds may continue to increase with an increase of frequency.

The proposed swimming motion can coordinate anterior, mid-body and posterior displacements, and is able to reduce the large kinematic errors seen in existing free swimming robotic fish. *iSplash*-I achieved a maximum velocity of 3.4BL/s and consistently achieved a velocity of 2.8BL/s at 6.6Hz with a low energy consumption of 7.68W.

Future experimental analysis will greatly benefit from visualization techniques accurately measuring fluid flow. However, a few assumptions can be deduced to account for the increase in speed: (i) The magnitude of propulsive force was increased by initiating the starting moment of added mass upstream; (ii) The developed structural arrangement allowed for smooth transition of flow along the length of body; (iii) Anterior and/or mid-body vortices were formed, coordinated and propagated downstream; (iv) Lateral and thrust forces were optimized around the center of mass; (v) A reduction in drag resistance due to reduced anterior amplitude errors.

We also aim to improve the performance of *iSplash*-I by applying higher frequencies, further optimizing of kinematic parameters along the full length of the body and the development of a prototype carrying its own power supply within a comparable sized frame.

ACKNOWLEDGMENTS

Our thanks go to Richard Clapham senior for his constant technical assistance towards the project.

REFERENCES

[1] P. R. Bandyopadhyay, "Maneuvering hydrodynamics of fish and small underwater vehicles," Integr. Comparative Biol., vol. 42, no. 1, pp. 102–17, 2002.

[2] J. Lighthill, "Mathematical Biofluiddynamics," Society for Industrial and Applied Mathematics, Philadelphia, 1975.

[3] J. J. Videler, "Fish Swimming," Chapman and Hall, London, 1993.

[4] G. S. Triantafyllou, M. S. Triantafyllou, and M. A. Grosenbauch, "Optimal thrust development in oscillating foils with application to fish propulsion," J. Fluids Struct., vol. 7, pp. 205–224, 1993.

[5] D. S. Barrett, M. S. Triantafyllou, D. K. P. Yue, M. A. Grosenbaugh, and M. J. Wolfgang, "Drag reduction in fish-like locomotion," J. Fluid Mech, vol. 392, pp. 183–212, 1999.

[6] J. Yu, M. Tan, S. Wang and E. Chen. "Development of a biomimetic robotic fish and its control algorithm," IEEE Trans. Syst., Man Cybern. B, Cybern, 34(4): 1798-1810, 2004.

[7] J. Liu and H. Hu, "Biological Inspiration: From Carangiform fish to multi-Joint robotic fish," Journal of Bionic Engineering, vol. 7, pp. 35–48, 2010.

[8] P. Valdivia y Alvarado, and K. Youcef-Toumi, "Modeling and design methodology for an efficient underwater propulsion system," Proc. IASTED International conference on Robotics and Applications, Salzburg 2003.

[9] R. Bainbridge, "The Speed Of Swimming Of Fish As Related To Size And To The Frequency And Amplitude Of The Tail Beat," J Exp Biol 35:109–133, 1957.

[10] M. Nagai. "Thinking Fluid Dynamics with Dolphins," Ohmsha, LTD, Japan, 1999.

[11] P. W. Webb, "Form and function in fish swimming," *Sci. Amer.*, vol. 251, pp.58–68, 1984.

[12] M. W. Rosen, "Water flow about a swimming fish," China Lake, CA, US Naval Ordnance Test Station TP 2298, p. 96, 1959.

[13] M.J. Wolfgang, J.M. Anderson, M.A. Grosenbaugh, D.K. Yue and M.S. Triantafyllou, "Near-body flow dynamics in swimming fish," September 1, J Exp Biol 202, 2303-2327, 1999.

[14] G. V. Lauder and E. G. Drucker, "Morphology and Experimental Hydrodynamics of Fish Control Surfaces," IEEE J. Oceanic Eng., Vol. 29, Pp. 556–571, July 2004.

iSplash Robotics

Published by *iSplash* Robotics UK
www.isplash-robotics.com
Copyright © 2016

Published by *iSplash* Robotics 2016
Illustration by Richard James Clapham PhD

All rights reserved. No part of this publication may be reproduced, stored in a retrieval system or transmitted in any form or by any means, electronic, mechanical, photocopying or otherwise without the prior permission of Natural Classics.

ISBN:
Thank you for your purchase.

www.ingramcontent.com/pod-product-compliance
Lightning Source LLC
Chambersburg PA
CBHW070342190526
45169CB00005B/2007